A nostalgic look at

CARS

A nostalgic look at

CARS

Lynda Springate

**PICTURES FROM
THE NATIONAL MOTOR MUSEUM**

Silver Link Publishing Ltd

First published in September 1996

British Library Cataloguing in Publication Data

A catalogue record for this book is available from the British Library.

Hardback ISBN 1 85794 093 8
Paperback ISBN 1 85794 101 2

Silver Link Publishing Ltd
Unit 5
Home Farm Close
Church Street
Wadenhoe
Peterborough PE8 5TE
Tel (01832) 720440
Fax (01832) 720531
e-mail: pete@slinkp-p.demon.co.uk

Printed and bound in Great Britain

All photographs are from the collection of the National Motor Museum, Beaulieu, Hants SO4 7ZN.

Riley is a registered trademark of Rover Group, Bickenhill, Birmingham.

BIBLIOGRAPHY

Birmingham, Dr A. T. *The Production and Competition History of the Pre-1939 Riley Motor Car* (Foulis, 1974)
Reeves, Edward *The Riley Romance* (Riley (Coventry) Ltd, 1930)
Styles, Dr D. *As Old as the Industry* (Author, 1982)
Taylor, James *The Riley RM Series* (MRP, 1990)
Riley Sales Brochures
Autocar road tests and announcements
Motor road tests and announcements

ACKNOWLEDGEMENTS

I would like to thank the staff of the Photographic and Reference libraries at the National Motor Museum, Beaulieu. Both libraries offer an invaluable service, with access to contemporary photographs, sales brochures, motoring journals, books and literature.

I would also like to thank my husband Frank for supporting me so enthusiastically throughout the compilation of this book, and for agreeing to invest in another Riley RM one day!

CONTENTS

Introduction   7

1.   **Pre-war Rileys**   9
2.   **Post-war Rileys**   26
     The RM Series   27
     Riley-based Healeys   55

Rileys in colour   57

3.   **The BMC years**   61
     Pathfinder   62
     Two-Point-Six   67
     One-Point-Five   70
     4/68   77
     4/72   80
     Elf   86
     Kestrel   94

Index   100

INTRODUCTION

Having built its first car in 1898, Riley could truly claim as its motto 'As Old as the Industry. As Modern as the Hour'.

The Riley family originally owned a weaving business based in Coventry, but in 1890 William Riley decided that there was more money to be made in bicycles and bought the cycle firm Bonnick & Co. The name was changed in 1896 to the Riley Cycle Co.

William had five sons. Percy, the engineering genius behind the cars, was to die in 1941 aged 59. Of the other four, Victor died in 1955, Allan in 1963, and Stanley and Cecil in 1961 and 1962 respectively. Victor joined his father in the family business when he left school, and Percy followed suit in 1896. Although Percy was very interested in motorised transport, William was firmly against it and Victor was too interested in building bicycles.

However, by 1898 Percy had designed and built his first car, which was tested by driving it to Stratford-upon-Avon. The car no longer exists, but it is known that Percy was among the first to produce an engine fitted with a mechanically operated inlet valve, which was to prove useful to the budding British motor car industry. In a subsequent court case Riley was able to prove that it had anticipated Benz, who had patented the design and wanted to collect royalties from it.

In 1902 Victor, Percy and Allan joined forces, and with money borrowed from their mother formed the Riley Engine Co. They were soon supplying air-cooled engines to Singer and the Riley Cycle Co for motor-tricycles. The following year, 1903, saw the first Forecar, with two wheels at the front and a single one at the rear. It could carry one passenger, who sat between the two wheels at the front. The company produced its first four-wheeled 9 hp car in 1905, which sold for £167.

Riley was by no means the first automobile manufac-

turer to begin life as a bicycle manufacturer. William Morris, later to become Lord Nuffield, was a keen cyclist. Humber began producing bicycles in 1868, and Singer and Triumph also followed this route. French cycle-maker Peugeot, on the other hand, manufactured items such as corset stays before diversifying into bicycles and cars.

During the early years of the century roads were little more than trackways and punctures were commonplace. The average motorist could expect to travel only about 8 or 9 miles before having to mend a tyre. A vulcanising kit was therefore an indispensable accessory, as repairs had to be carried out on the spot. Percy Riley's answer to this problem, so obvious today, was to

William Riley (1851-1944), who took over management of the family weaving business in 1870. He had the foresight to realise that the Education Acts of the time would bring about the decline of cheap child labour in factories, so he exchanged his weaving business for cycle manufacture, ensuring his sons' keen interest in road transport. He outlived his engineering genius son Percy by three years, and his wife had died at the early age of 55 in 1909.

THE FIRST RILEY CAR DESIGNED & BUILT BY Mᴿ PERCY RILEY 1896-1898
This was the first car in which a mechanically operated inlet valve was
incorporated. This feature was copied in 1899 by the Benz Co. of Germany.

352

Percy Riley worked on his first car design between the years 1896 and 1898. Much of the work had to be carried out secretly, as his father was too pre-occupied with bicycle manufacture to be interested in cars. It was a little two-seater with a floor-mounted steering wheel, and had what was commonly known as a 'coal scuttle' bonnet. Its most notable feature was a mechanically operated inlet valve. When the German firm of Benz tried to collect royalties on the design, Percy was able to prove that he had already been working on the design. Successful trial runs were carried out in the machine, which now no longer exists.

design detachable wheels, and the 9 hp Riley of 1907 was the first car to be thus equipped as standard. Wheel production soon be came a highly profitable arm of the Riley organisation, and they were also fitted to many racing cars.

Pleased by the success and profitably of detachable wheels, and having decided that there might after all be a future in the automobile, William gave up all opposition and cycle manufacture was discontinued in 1908. By 1912, however, he was again opposing car production in favour of concentrating wholly on detachable wheels. The company name was then changed to Riley (Coventry) Ltd, and it was left to the brothers to take over the car-making interests, forming the Riley Motor Manufacturing Company. Their first new model of any note was the four-cylinder 17 hp.

During the First World War all five Riley brothers volunteered for active service. However, Victor, Stanley, Percy and Allan were told to remain at the works and turn the factory over to war production. Cecil returned from the war a Captain in the Royal Flying Corps.

In 1919 Percy and Stanley Riley purchased the plant and machinery of the old Nero Engine Company in Durbar Avenue, Foleshill, Coventry. Riley (Coventry) Ltd had by this time given up making detachable wheels, and the Riley Motor Manufacturing Co had become the Midland Motor Body Co. In 1931, an eventful year in the Riley family history, Riley (Coventry) Ltd took over the Midland Motor Body Co and Riley Engine Co. It was the Foleshill site that was to produce some of the best-loved pre-war Rileys until production was moved to the MG factory in Abingdon in 1948.

A Riley Forecar, pictured here from the rear, was driven by Victor Riley in the 1934 London-Brighton Emancipation Run. Riley produced its first three-wheeled 3 hp Forecar in 1903. The model continued in production until 1904, by which time it had acquired a larger 4½ hp engine. The driver piloted the car from behind the passenger, whose seat was suspended between the two front wheels.

1. PRE-WAR RILEYS

Percy Riley's most famous design was the Riley Nine, which first appeared as the fabric-bodied Monaco saloon in 1926. Although a small saloon car, 12 ft 5 in long, the Nine was an instant success as it combined elegant styling with good performance and power. It was capable of a top speed of 60 mph and was fitted with what would be known today as a cross-flow cylinder head.

Despite its quality of finish and advanced features, the Nine was very firmly aimed at the middle range of the car-buying market. Stanley Riley is credited with the design of the coachwork, which with its low roof-line and outside luggage carrier was less box-like in shape than other small cars of the period.

The Monaco was not intended as a sporting saloon, but J. G. Parry-Thomas and Reid Railton modified a Riley Nine engine for racing. Unfortunately, Parry-Thomas was killed in 1927 while attempting the World Land Speed Record in 'Babs'. Railton persevered and produced the first 'Brooklands' Riley, which was capable of 80 mph. It was during that year that Victor Riley became Managing Director of Riley (Coventry) Ltd.

The Nine was to achieve numerous competition successes in the hands of professional and amateur drivers. Noted driver George Eyston lapped the Montlhery circuit in France for 48 hours in a Monaco, creating a new 5,000 km record. A specially equipped Monaco entered the 1931 Monte Carlo Rally; starting in Norway, the car won the 1,100 cc Class. Donald Healey entered a Riley Nine tourer in the Paris-Nice trial of the same year and also won the 1100 cc Class.

Contemporary car magazines of the period would often carry reports of hill-climbs and long-distance trials, organised by the Motor Cycle Club or RAC; favourite events were the Welsh Rally, Scottish Rally and Land's End Trial. The advantage of this peculiarly British motor sport was that it appealed to all pockets.

Events were organised fairly and into classes, so that each competitor stood a chance of winning something. Popular choices of car were MG, SS Jaguar, Alvis and Singer, and nearly any magazine report would have at least one mention of a Riley competing, more often than not successfully.

An enlarged version of the Nine engine was the six-cylinder 14/6 of 1929, which was in production for five years. In 1932 the 1,458 cc six-cylinder 12/6 was introduced, while the 15/6 of 1935 had a 1,725 cc engine and was fitted with a pre-selector gearbox as standard. Four-cylinder and six-cylinder Riley chassis were able to be fitted with virtually the same styles of bodywork, and a number of different body styles were soon offered for both. One of the company's 1933 sales brochures, 'The Book of the Riley Nine', catalogued the Monaco, Kestrel, Falcon, Mentone, Alpine and Stelvio saloons, Ascot and Lincock coupés, and Edinburgh and Winchester luxury saloons; the latter two came equipped with extras such as 'two ashtrays in the front, ladies' and gentlemen's companions, two folding tables, and silk blinds to windows'. A specially bodied sports, the March Special, was also listed.

Many people argue that Riley produced two of the most beautiful sports cars of the 1930s, the Imp and the

The front cover of the sales brochure 'The Book of the Riley Nine', showing the 'Such fun to drive' slogan and famous blue triangle with 'As Old as the Industry, As Modern as the Hour' in the white surround. Riley introduced its Plus Ultra range for 1932. In this series the body construction was given a new dropped chassis frame, which gave occupants an increase in headroom without spoiling the lines of the car.

larger MPH, while the Kestrel fastback saloon version, also fitted to the larger models, was ahead of its time with its sloping rear styling.

These days the idea of audible warning systems or on-board computers is taken for granted, but in the early 1930s Riley piloted an early-warning system on its six-cylinder models. If the mixture strength control was left on the 'rich' setting, a warning was sounded when the driver used the horn button.

One of the most important Rileys of the mid-1930s was the 1½ litre, designed by Hugh Rose, who later moved on to Lea Francis and created a very similar engine. The four-cylinder 1,496 cc unit was to stay in production for close on 20 years. Unusually for a British firm, Riley produced a V8 engine in 1936; a 2,178 cc unit, it was available as the Adelphi saloon or Kestrel saloon.

One of Riley's best-known main agents of the time was the firm of Jimmy James Ltd, based in Euston Road, London. They had already been distributors for 11 years when in 1938 they were appointed sole distributors for London and Essex. An article in *Autocar* in that year states that a new service station at Highgate 'will be able to hold 250 cars and will be devoted to the servicing of Riley cars'.

Despite maintaining a high reputation for looks, handling, comfort and sheer enjoyment of driving, it has to be said that Riley's extensive range of models and body styles must have contributed substantially to the company's financial downfall in 1938. A subsidiary company named Autovia had been set up in 1937 to produce luxury cars utilising the Riley V8 engine, but the car was expensive at £975 and very few were sold. The last Riley to be produced before the company went into receivership was the appropriately named Victor, available with a choice of 1,087 cc or 1,496 cc engine.

Reports had appeared in the motoring magazines in 1937 of a proposed merger between Riley and the Triumph Motor Co. However, this was not to be, and early March 1938 saw Victor Riley make a statement to the press:

'The directors have given long and anxious consideration to the financial position of the company, and in view of the difficulties experienced in carrying on the business they have felt constrained to ask the company's bankers to appoint a receiver to protect the assets in the interests of all concerned.'

Sir W. H. Peat was duly appointed receiver on 24 February. In October Lord Nuffield bought Riley for £143,000 out of his own money, then sold it on to Morris Motors for the rumoured price of £1. Many Riley enthusiasts felt that this was the end for the 'thoroughbred Riley', but Lord Nuffield announced that the company would be left to run its own affairs, and Victor Riley was to remain Managing Director of Riley (Coventry) Successors. However, the model ranges were immediately trimmed down, and 1939's cars were developments of the Big Four and 1½ litre cars. The 12 hp in saloon form was priced at £310, the 16 hp at £385.

The Second World War put paid to any further developments for the time being at the Foleshill works, as the factory was turned over to war work. As is well known, Coventry was one of many key British cities to suffer saturation bombing by the German air force. One of the casualties was the Riley works, which lost the body jigs for the latest designs in an air raid.

Marketed as the 'Wonder Car', the Riley Nine first appeared in 1926 as the Monaco saloon and soon established the Riley family as makers of reliably engined cars, capable of good performance and handling. Percy Riley designed the engine and Stanley the bodywork, and the first Monaco went on sale priced at £285. Riley also offered a colonial version, which had a higher chassis. Various models of the 9 hp spanned the years 1926-36. The car featured here is a 1932 Riley Nine Monaco, which would have cost £298.

A close relation of the Riley Nine was the 14/6, introduced in 1928. The engine was a small six-cylinder 1,633 cc version of the Nine engine, with a chassis 6 inches longer. It sold for £495 in saloon form, and was subject to £14 per year annual horsepower tax, as opposed to the Nine's £9. The 14/6 enjoyed a wide choice of body styles. The sales catalogue quoted the Stelvio, Alpine, Kestrel, Edinburgh, Winchester and Deauville saloons, the Special, Alpine and Lynx tourers, the Sportsman's coupé, Lincock fixed-head coupé, Ascot drophead coupé, and two limousine options. The occupants of this 14/6 saloon are enjoying a visit to Penshurst in Kent, still a popular tourist venue famous for its historic house and garden.

Left The dependable engine and handling capabilities of the Nine made it an inevitable and enviable choice in British endurance trials. There were many such events in the 1930s and 1950s, and some are still continued. This illustration shows a 1927 9 hp tackling the MCC Land's End trial at Barbrook Mill, near Lynton, North Devon.

Below left In many ways Great Britain was ideal for trialing. Its relatively small size meant that trial organisers had a variety of terrains that could be covered fairly quickly by competitors. It was also very popular as a spectator sport, and there was never a shortage of people to give a recalcitrant car a shove if need be. Here a Riley Nine picks its way along a course lined with observers on a very damp day.

Above One of the major trial organisers was the Motor Cycle Club (MCC). This well-established body originally organised motor cycle events, but later it was natural that it should add car events to its programmes. Key long-distance events were the London-Exeter, London-Land's End and London-Edinburgh. Negotiating a tight corner on the latter is car No 147, a fabric-bodied Riley Nine Monaco of 1929.

Oldest of the Riley Clubs is the Riley Motor Club, formed in 1925. In 1958 the club literature promotes activities such as tours, hill-climbs, rallies and social runs in addition to social events such as visits and dinner dances. Membership at this time cost £2 2s 0d per year, including RAC affiliation. The Vice-Presidents at that time included such famous names as the Duke of Richmond and Gordon, Earl Howe and S. C. H. Davis.

OFFICERS

President

Vice-Presidents   His Grace, The Duke of Richmond and Gordon
The Rt. Hon. The Earl Howe, P.C., C.B.E., V.D.
The Rt. Hon. The Lord Mostyn
S. C. H. Davis, Esq.
J. R. Woodcock, Esq., O.B.E.
F. R. Gerard, Esq.

Hon. General Secretary and Treasurer   *A. Farrar, Esq.*

T HE Riley Motor Club was formed more than twenty-five years ago to serve nationally and through locally organized and locally run Centres the interests of Riley car-owners, their families and friends.

The main object of the Club is, naturally, the promotion of motoring activities—tours, hill-climbs, rallies, social runs, etc., but attention is also given to the arrangement of purely social events—dances, dinners, and the like.

Notice of these occasions, news of the Club and its local Centres, and much that is of general motoring interest are given to members in the pages of their own monthly magazine, *Motoring* (this publication is the official organ of the Riley Motor Club: it forms part of the Club service).

In addition to the above the Club does its best to secure special facilities (such as member car parks) at the main motoring events.

R.A.C. affiliation, entitling members to all the benefits of R.A.C. Associate Membership, may also be obtained through the Club.

The Riley Motor Club enjoyed many rallies and visits, including these to Croydon Aerodrome (*above left*) and the Kodak factory at Harrow (*left*). Note that the soft top of the car in the foreground at Croydon incorporates the Riley diamond!

Above American car designers of the mid-1930s began experimenting with wind-cheating 'streamlined' designs. One of the earliest was the ugly and ill-fated Chrysler Airflow of 1934, which was hurriedly withdrawn. Riley, however, had designed the 'fastback' shape of the Kestrel saloon by 1932. With its almost diamond-shaped windows and sloping rear-end treatment, many people regard it as one of the most classic and successful fastback designs. The style was available on the 12/6, 9 hp, 14/6, 15/6, 1½ litre and Big Four. There is a very slight possibility that the failed 8-90 may also have had a Kestrel version.

Right After the original 'Brooklands' Riley was produced by Parry-Thomas, Riley began to offer a sports-racing Brooklands model from 1929. The short-lived Brooklands Six was offered from 1932. As with the saloons in long-distance rallies, the racing models did well as competition cars.

In 1932 a Riley team headed by Captain George Eyston (later to be known as the driver of the record-breaking MG Midget cars) attempted a number of class records on the Montlhery track in France. He averaged 92.82 mph over 12 hours and 91.68 mph over 2,000 km.

Left Tuning wizard Freddie Dixon began racing cars in 1932, and his previous experience in racing and tuning motor cycles was to prove a great advantage. He entered his own Riley Nine Brooklands in the 1932 TT race; the car was fitted with a lightweight streamlined body and proved extremely fast. Unfortunately Dixon lost control of it at Quarry Corner and ended up in a rhubarb patch. Freddie's most successful year was in 1935 when he won the TT and British Empire Trophy.

Below One of Freddie's best-known cars was the Riley-based Special known as the 'Red Mongrel'. It was a single-seater with a modified Riley TT engine using four carburettors. It scored numerous class record wins, but was not a particularly reliable car and Dixon subsequently sold it.

Right Racing driver Raymond Mays was an enthusiastic Riley owner who drove a Riley Nine Biarritz. In an open letter to the company he wrote, 'I fully understand the great enthusiasm of all Riley 9 owners.' Mays won many hill-climbs and raced at Crystal Palace and on the famous Brooklands circuit. Later he campaigned successfully in his own Riley special, the 'White Riley'. It was from this car that the racing marque ERA (English Racing Automobile), founded by Mays, Peter Berthon and Humphrey Cook, was developed. Mays raced the ERA in the light-car racing classes, and founded BRM after the Second World War.

Below The 'White Riley' sprinting up Shelsey Walsh. The car's engine, with modifications designed by Peter Berthon, was a six-cylinder 1,486 cc unit fitted with a supercharger. The car is still in existence.

Above A team of Brooklands Riley Nines was entered in the 1929 Ulster TT. The cars being checked over here were driven by W. P. Noble, J. R. Cobb, S. C. H. Davis, R. S. Outlaw/V. Gillow, and E. Maclure/P. Maclure. Riley took the first three places in the 1,100 cc Class.

Left Two Brooklands Riley Nines showing their manoeuvrability at Brooklands in front of an advertisement for *The Motor Cycle*. The track was created in 1907 by Hugh Locke King because the laws in Britain at that time made no allowances for road racing. Nearly every great name in motor sport raced there; some, like Sir Malcolm Campbell, even had workshops. Many endurance record attempts took place at Brooklands, such as the JCC Double Twelve, when cars were raced for 12 hours, locked away in an enclosure, then raced for another 12 hours.

Brooklands race track was closed down for the duration of the Second World War, and sadly never re-opened. The sheer enthusiasm and vitality of the crowd and drivers can only be imagined today, although some of that atmosphere still survives in and around the clubhouse, which is now a museum.

Right Another British racing hero to start his career by racing Rileys was Mike Hawthorn, seen here driving a Riley Imp around Goodwood. Mike Hawthorn began racing with his father in 1950. They each drove a Riley, and in 1951 Mike won the Ulster and Leinster trophies in Ireland, as well as the Motor Sport Brooklands Memorial Trophy for club racing. He very quickly went on to race successfully for Ferrari, but had an unpleasant time being harried unjustly by the British press and some MPs over his National Service. In 1955 Mike joined the Jaguar car company driving a D-type, then, after a short period of driving for BRM, he rejoined Ferrari and stayed with them until he decided to retire. Having become World Champion at the age of 29, tragically he was killed not on the race track, but on the Hog's Back near Guildford, when his car hit a tree.

Below A nerve-racking test in front of a crowd of spectators during the 1934 RAC Rally. The Riley MPH seems to be coping well, with the help of the passenger.

Above New for 1936 was the four-cylinder 1,496 cc Riley Sprite, which was to be the last true sports two-seater produced by the company. It was heavier in style than the Imp and MPH, but still an attractive motor car, capable of 85 mph. The driver of this Sprite, taking part in the Great West Trial in 1938, seems to be demonstrating the 'hand-brake on, gently turn the wheel slowly' technique of getting out of ditches (with a little help).

Left D. R. Lloyd's Riley Sprite is seen here fencing with a pile of sand-bags. The angle of the photograph shows off the raked-back 'fencer's mask' grille of the car.

It seems to have been much easier to use an everyday car competitively in the 1930s. It would not be so easy to cordon off a seafront such as that at Brighton today, when far more parking is required by visitors to tourist attractions. Note the 30 mph sign on the ornate lamp-post.

Right A Lynx tourer in the 1937 Welsh Rally enjoying the scenery; its entry card shows that it started from Norwich. Many of the awards for this Rally went to the SS car company, but Riley owners still put up a good show. J. F. A. Clough's Riley won a starting control prize, and Mrs Hague won the ladies' prize for open cars and the ladies' eliminating prize. Reporting on the eliminating test, *Autocar* announced that 'J. F. A. Clough hurled his grey Riley Sprite round in 1 minute 6⅘ seconds'.

Below A Kestrel taking part in an uphill restart test. In those days spectators and drivers were all smartly dressed, whatever the weather conditions. Smartest of all is the little girl in the hat being held by her mother, next to the marker pole on the right.

The Women's Association of Sporting Amateurs was formed to cater for the ever-growing number of talented women drivers, and organised a great many trials and rallies. This picture was taken at a 1938 event, and the interesting cars in the line-up next to the Riley Sports two-seater are an OM Tipo 665, a Railton Straight 8, a BMW and an MG.

Above Riley's 'white elephant' could be said to have been the Autovia car company, set up by them in 1937. Having developed its V8 engine, Riley set about trying to produce its own luxury car aimed at the Bentley saloon or small Rolls-Royce buyer. The car was described by *Motor* as 'an entirely new eight-cylinder car', which went on to praise its quiet running and riding comfort. The new car also impressed the judges at the Ramsgate Concours d'Elegance, where it won first prize in the class for cars costing up to £1,000. Unfortunately its price of £975 was far too high in an already overcrowded market. Only a few were made and production ceased in 1938. Many years later, in 1951, in a letter to *Autocar*, a Mr W. Bretherton wrote, 'I am the owner of one of these excellent cars, it being a seven-seater limousine first registered in 1940. The Autovia car company of Coventry produced a small number of these cars. . . The Autovia is no longer in production (more's the pity), the reason I am told being that they proved too costly to make.'

Right The 'Riley Ski Lady' mascot, designed for the radiator of the Light Six when it became known as the Alpine, and the 'Kestrel' mascot for the range of the same name. Only a few 'Ski Lady' 'factory' mascots,

those specifically associated with this particular marque of car, were produced. Copies were, however, provided by other companies who produced all kinds of designs of radiator mascot; many owners bought 'accessory' mascots from outlets such as Gammages, or even from a shop that is still very much with us today, Halfords!

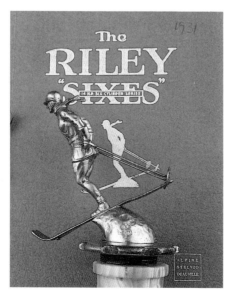

2. POST-WAR RILEYS

It took the motor industry some time to get back on its feet again after the war. For a time most cars represented a continuation of those produced in 1939, and this can sometimes make dating a problem for enthusiasts restoring cars of that period today. Not only car manufacturing was affected. This was the 'Utility' era, and everything was in short supply. As a result manufacturers were rationed as to the amount of materials they could use. In the case of wooden furniture only two utility styles were available - quality was good but very simple in an effort to conserve wood.

Rationing was still the norm. Luxuries such as face-

cream and lipstick were few and far between - if women saw a queue outside of a chemist's shop they used to join it regardless, as it meant fresh supplies were in. Even soap and soap powder were rationed, and sometimes shaving soap was grated into the weekly wash to eke out supplies. The Black Market was still flourishing, and the police devised a check that showed if a car was running on illegal petrol or not. To make matters worse for Britain, the winter of 1947 was one of the severest on record. There were restrictions on allocation of steel for saloon cars, unless they were for export or farm work; it was this latter restriction that inspired Rover to pro-

This new Riley has a clean, graceful appearance from whatever angle it is viewed. Clever design reduces wind resistance and simplifies cleaning.

The windscreen is V-shaped, opening on the driver's side. Instruments are mounted centrally on a polished walnut facia board. Steering wheel reach is adjustable by a telescopic steering column.

NEW OUTSTANDING RILEY

Accommodation provided at the rear of the car for luggage is unusually spacious. The luggage boot is sufficiently deep to house, unobtrusively, a surprising quantity of luggage with complete protection from weather.

duce the Land-Rover. The Standard Vanguard of 1947 was the first British car to have modern post-war styling based on the current American trend.

It was in this climate that the first of the new Riley RM Series, the 1½ litre RMA, made its debut in September 1945, powered by a re-developed version of the pre-war 1,496 cc engine. The frame was still made in the traditional way of ash, then fitted with steel panels. In some ways it was very brave of the Nuffield organisation to offer such a well-finished, stylish sporting saloon at this time. The RMA was aimed at the middle section of the car-buying market, but priced in 1946 at £675, or £863 with Purchase Tax, it was not a cheap car - a Ford Prefect in the same year cost only £352, while a Bentley cost at least £2,997. Cost on the domestic market was, however, to be immaterial for some time, and the Government's 'Export or Die' policy accounted for many of the new cars. Other problems were caused by labour unrest in Coventry, the inability of suppliers to deliver parts, and the Government's system of issuing permits to buy cars; in 1945 Riley (and other manufacturers) had to honour their previous commitments before cars could go on sale to the general public.

In many ways the Second World War bought about a clean sweep for the Riley company. While the 1939 and 1940 Rileys were very similar in appearance to the other Nuffield marques of Morris and Wolseley, the RM Series was altogether different, and marked a return to the company's roots. Its styling was not dissimilar to the elegant Citroen Traction Avant or the German BMW, and would certainly not be confused with any other Nuffield product.

It was not only looks that the RMA shared with the Traction Avant - Riley engineers had examined the suspension of the Citroen and were impressed by it. Although the French car had front-wheel drive, while Riley used rear-wheel drive, similarities were incorporated into their own torsion bar suspension system. In addition to independent front suspension, the new Riley had rack-and-pinion steering, which much improved the handling. Riley's Chief Engineer was Harry Rush, and it was his team that was responsible for the new car. Sadly he was killed one night driving a 2½ litre Riley; this was after production had moved from Coventry to Abingdon, and he was travelling between the two locations.

Production of the 2½ litre RMB followed in 1946. This time the engine was the earlier Big Four unit of 2,443 cc. Otherwise the RMB was mechanically the same as the RMA, but 7 inches longer. From the front the two models can be quickly identified by the colour of the radiator badge, dark blue for the 1½ litre and light blue for the 2½ litre.

Riley cars did not sell particularly well in the United States, certainly in nothing like the numbers of contemporary Jaguars and MGs. In 1948 the Roadster was produced very much with export in mind; offered as a three-seater with column shift, a two-seater with floor change was later offered for home consumption. Even so, neither the Roadster nor the Drophead Coupé sold in any great numbers and were taken out of production after only three years.

It is interesting to note that at this period the cars were simply known as the 1½ litre or 2½ litre, the 'RM' designations not being applied until 1952 on the updated models, the older cars being given the type codes 'RMA' and 'RMB' in retrospect.

he rear seat is exceptionally wide ith 51 in. (129.53 cm.) effective idth ; there is ample room for ree people and deep wells provide enty of leg room.

pholstery is softly sprung and is immed in best quality English ather.

ATURES

f the " Torsionic showing swivel struts, dampers mechanism.

and axle are ated to ensure

Although Riley wasn't the first manufacturer to start producing cars again after the war's end, it was one of the first to offer a completely new design. The 1½ litre, which was later to be known as the RMA, went into production for 1946. It was also the first British car to feature rack-and-pinion steering. This early 1946 sales brochure shows some of the new car's main selling points. The dashboard, although simple, was finished with a walnut veneer, and much was also made of the luggage space with the spare wheel in a special housing separate to the rest of the boot. 'Torsionic suspension' was of course made much of. The sales brochure went on to report that 'Wheel changing is clean and easy, a task capable of being performed by the lady driver'. Equally impressive was a top gear of 75 mph with a fuel consumption of 30 mpg.

1½ LITRE 2½ LITRE

2½ LITRE SALOON (in Silver Streak Grey)

—and here's the secret of super performance

Among those motorists who study engine details the Riley has always held a high position of esteem because enthusiasm for the niceties of engine design is often the first step towards Riley ownership. The world-famous Riley is available in a 2½ litre model, capable of 100 miles an hour, and a 1½ litre model which also provides the thrill of sports car performance. Proved leadership in design makes permanent a Riley owner's enthusiasm for his car.

FOR MAGNIFICENT MOTORING

Left 'Magnificent Motoring' is summed up very nicely in this 1½ and 2½ litre brochure. 'Silver Streak Grey' was a colour introduced for the 1953 model year, and was presumably thought to be a more interesting description than the simple 'grey' that it replaced. Metallic paint was available as an option from late 1951, known as Metallichrome. Marketing was aimed at the businessman who preferred to drive himself.

Right Although owned by the Nuffield company, Riley cars of this period bore no resemblance to any other Nuffield product. Eric Holmes and Eric Carter were responsible for the car's elegant lines. The roof was finished in vinyl, which gave a pleasing effect and may also have helped save steel supplies. Supply shortages and the Labour Government's export policy meant that very few people had the opportunity to buy the new car for some time.

In 1952 the 1½ litre cost £750 and the 2½ litre £958; top speed of the latter was nearer 94 than 100 mph, although this was still a respectable top speed for a saloon car weighing 34½ cwt. The four-cylinder 2,443 cc engine had a power output of 100 bhp at 4,400 rpm from October 1948, slightly more than the previous 90 bhp.

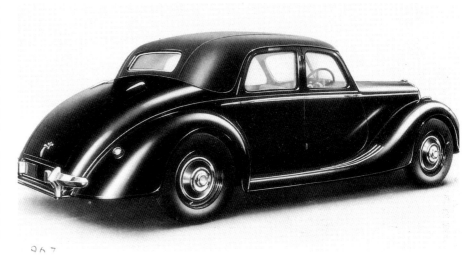

Below This 1948 sales brochure shows the 2½ litre Drophead Coupé - note the Nuffield Exports badge at the bottom. That year saw a return to something like normality for the British motor industry. The Society of Motor Manufacturers and Traders held the first London Motor Show since the war, and several future 'classic' cars were on display there for the first time, among which were the Morris Minor and Jaguar XK120. Riley showed their Drophead Coupé at the '48 show for the first time, although it did not actually go into production until 1949. Only 502 Dropheads were made, and production was phased out in 1951.

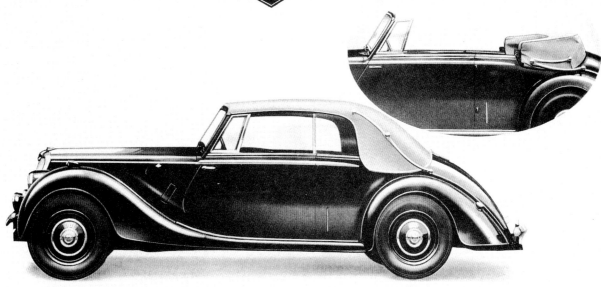

PUBLICATION No. NEL 113

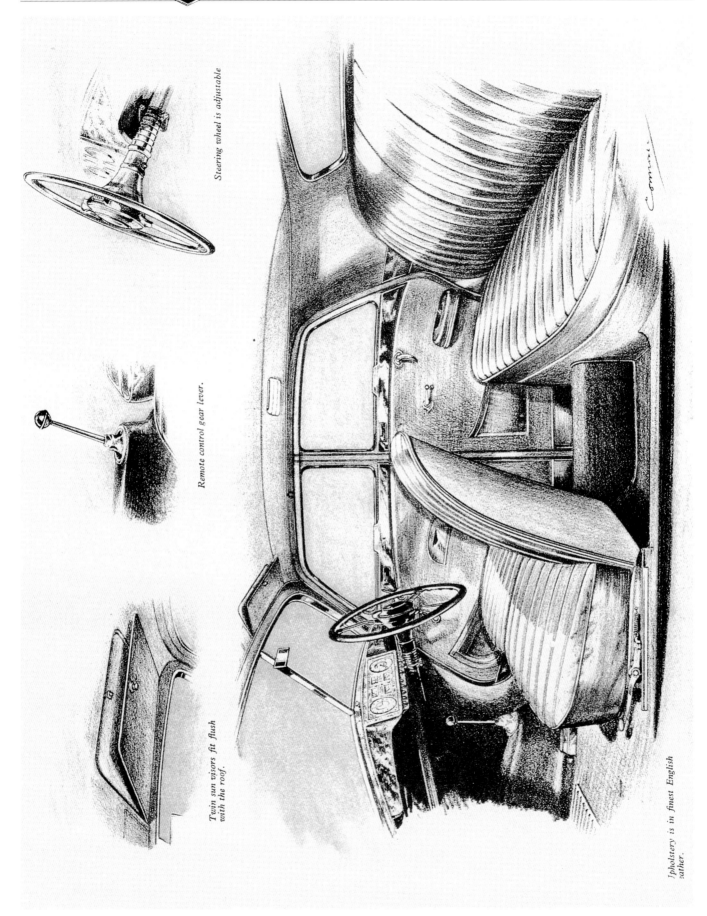

Steering wheel is adjustable

Remote control gear lever.

Twin sun visors fit flush
with the roof.

Upholstery is in finest English
leather.

The Riley 1½ and 2½ litre cars differed only slightly. The wheelbase of the 2½ litre was 9 ft 11 in, while the 1½ litre had a wheelbase of 9 ft 4½ in. The larger model also had twin SU carburettors. However, the interiors of both models displayed the same attention to detail, with such niceties as an adjustable telescopic steering wheel and an HMV car radio as a standard option.

The Drophead Coupé (*below*) was equally well finished, and had a quarter-light rear window. The hood mechanism was fixed just above the windscreen, so could be easily operated from inside the car. The Drophead shared the same basic dimensions with the 2½ litre saloon, and was only available in the 2½ litre capacity.

Interior of the Drophead Coupé is luxuriously appointed.

Radio is easily fitted

Retractable hood is quickly released. (*Coupé*)

Easily regulated ventilation. (*Coupé*)

Left The illustration from the front cover of a sales brochure for the Riley 2½ litre. Beautifully drawn, it very simply yet effectively represents something of the traditionally British character associated with Riley.

Below Designed to appeal to those who bought their cars to enjoy them, this 1951 Riley certainly seems destined for the 'open road'. Obtaining petrol in Great Britain was, however, still a problem; although ten years of petrol rationing was officially over, it would be a while before brand name petrol was back on the market, a situation not helped by Iran's attitude to British oil properties. Meanwhile, 'Pool' petrol had a tendency to make the engines of most cars 'pink'.

Above A fine example of artwork from the 1952 1½ litre and 2½ litre sales brochure, this time promoting the virtues of the Riley as a touring car. The caption reads, 'The high standard of performance, perfect riding comfort and exceptional luggage space of a Riley are shown to advantage on a holiday tour.'

Below The Riley 1½ litre RMA saloon was a heavy car at 25½ cwt, which may have accounted for the fairly slow 0-60 mph time of 25 seconds. However, top speed was 75 mph, with the car returning 28 mpg. Brakes were Girling Hydro-mechanical. Approximately 10,500 had been sold by the end of the car's production run in 1952, when it was superseded by the RME. Cosmetic differences between the two were most apparent in the re-styled front wings and lack of running boards on the RME.

Left Most of the Riley coachwork styles were produced at the factory. This Riley 1½ litre Drophead Coupé is unusual in that it was done by the private firm of coachbuilders, Bonallack. A limited number of these models went into production.

Below Taking part in the *Daily Express* Rally of November 1952 is this unusual Riley 1½ litre, boasting a wooden shooting brake-type estate body, not unlike the Ford Woody. Estates were not produced by Riley, so this would be a conversion carried out by a specialist coachbuilder.

Right Riley production moved from Coventry to Abingdon during 1948, and by 1949, the year of this 1½ litre saloon, production had started alongside the MG. The Nuffield company felt that it made more sense to have both sporting marques produced side by side.

Below In the days of the Channel Tunnel and Ro-Ro car ferries, it is easy to forget that travelling abroad nearly 50 years ago was not as easy as it is now. Many people judged it to be worth the effort, however, and took their cars touring. This 1½ litre proudly displays a GB plate as it is loaded aboard the MS *Kronsprins Frederik* at Harwich.

One of the most fashionable of advertising ploys has always been the posing of a car beside an aeroplane. Hillman, Morris and scores of other international manufacturers did it before the war, and Riley obviously decided to tackle the subject in the 1950s with this 1½ litre. Today we see photographs of Rolls-Royce and Jaguar cars shot in similar ways, and of course aircraft and car manufacturer Saab has its own version.

Presumably a posed picture to show the versatility of the Riley car and its owners.
This shot (no pun intended) is unusual in that the 2½ litre saloon is fitted with a radio aerial.

An early Riley 2½ litre parked outside of a country residence. Riley cars were owned by a number of well-known people, one of the most famous being Earl Mountbatten of Burma, who bought a 2½ litre soon after the car was introduced. He had the colour scheme finished to his own specification of Pale Blue and Black. The RMB also found favour with the police. A news item in *Autocar* of July 1948 reported that the Gloucestershire Police had ordered a flying squad of 2½ litre Rileys, equipped with radios.

Above It was not only the USA that saw Riley imports; this is Colonel B. B. Young of the Malaya Police Force with his 2½ litre Riley. A surprising number of European and American cars were exported to Malaya and Singapore, and a thriving classic car movement survives there.

Below The lines of the 2½ litre RMB seem even more elegant in a light colour. The split windscreen was to remain a part of Riley RM styling for the length of the model's life, the only exception being the Roadster version.

Above Riley revised the 2½ litre RMB model in late 1952. Re-named the RMF, one of the most important changes was the substitution of the traditional Riley spiral bevel back axle and torque tube transmission for an open two-piece propeller shaft and hypoid axle. All-hydraulic brakes was another of the changes.

HJB 622, pictured here, was a Riley press car and was the subject of an *Autocar Road* Test dated 26 December 1952. The reviewer, who tested the car in Belgium, reported, 'To those used to the extreme softness of some of the modern suspensions, the Riley may appear to provide a somewhat refined "vintage ride". Whilst it does not follow a transatlantic tendency towards sick-making softness, neither is it harsh.'

Left A view of the Riley hypoid axle as fitted to the updated RME and RMF models.

Above right The RME, an updated version of the RMA, also went into production in 1952. Like the facelifted RMB, the new model had an open propeller shaft and hydraulic brakes all round. There were also changes in the styling - the cars had no running boards and 'helmet'-type front wings. The rear window was also made larger.

Right Two-tone variations of the Riley RME look particularly elegant from the side view. The RME of 1954 was slightly slower than when it was first introduced - 0-60 took 31.8 seconds compared to 25.1.

Left The last RME saloon was produced in 1955, after a production run of 3,446. For the final two years of its life it had been one of only two Riley models to be offered, and was the last in the old traditional style. The RMB had been phased out in 1953 in favour of the new-style Pathfinder.

Below Leading a group of Lancia cars is this 1952 Riley 2½ litre, taking part in the 1954 Lancia Motor Club Rally. One-make car clubs would often organise road rallies that were open to members of other car clubs, and such events still take place today.

The Riley

100 h.p. 2½ litre open Sports 3-seater

"EVERYBODY who is wise on motoring matters knows by now that the 2½ litre Riley "goes like a bomb." In fact, the average speeds which they claim are getting very near to traditional fishermen's stories. There is no doubt, however, that this Riley is genuinely a very fast car, and an eminently roadworthy one as well."

—*vide "The Autocar."*

"THE lines of the new model are pure Riley. At the rear the boot is carried down in a graceful sweep of distinctly transatlantic suggestion, which at once provides luggage space of truly exceptional proportions and gives a very imposing appearance. The whole car, in fact, has an unusual air of sleek roominess that is entirely pleasing."

—*vide "The Motor."*

A new Roadster model with the 2½ litre engine went on sale in 1948, and was initially only available as an export model. The car's styling was very much aimed at the overseas market, in particular the United States. Column control on the early models gave the car enough room in the front to make it an occasional three-seater, while later models produced for sale on the home market had a floor change.

Above The Roadster was noted for having ample luggage space, as shown here. The scene is Morecambe, during the Lancashire Automobile Club's Morecambe National Jubilee Rally. Described as the Heat Wave Rally, it was hot on that May weekend in 1952 - so hot that many participating cars experienced overheating problems in the Lake District section. There were nearly 300 starters, and Morecambe Promenade was the finishing point. The Challenge Trophy was won by K. Bancroft in a Morgan, but a Riley was one of the class winners in the Concours d'Elegance.

Below A variation on the convertible theme was this unusual body style on a 1½ litre, believed to be by Epps Brothers of South London.

Post-war competition Rileys

Above Post-war Rileys never went back to the sort of competition results achieved in their pre-war years. Even so they were useful in endurance events like the Monte Carlo Rally. They were also popular in the numerous road trials that took place during the 1950's. Here the passengers on a Southdown bus look down on a 1½ litre saloon competing in the 1952 Eastbourne Rally.

Southdown buses were once a familiar sight along the South Coast from Sussex to Eastern Hampshire. This is one of a batch of six Leyland Cub KP3As delivered in 1936. The 20-seater coach body by Harrington provided plenty of space for the passengers and was popular for excursions and private hire.

Right A Riley 2½ litre Drophead Coupé passes two observers and their Sunbeam-Talbot. Dropheads were given a deeper windscreen header from 1950, as this example illustrates.

Two competitors in the 1954 Redex Rally are photographed high up on the Pennines. Turning the corner is a 2½ litre saloon, and waiting to join them a Citroen Traction Avant. There were certain styling similarities between the two marques, but while the Citroen had a monocoque body, the Riley still used a separate chassis. The Riley Torsionic front suspension was also very similar to the suspension system used by Citroen.

Above A 2½ litre Drophead manoeuvres around a marker with four up. Although there were plans for the smaller 1½ litre to be produced with a Drophead version, the cars were never put into production.

The Drophead Coupé began production in 1949 and lasted until 1951, during which time only 502 were manufactured.

Right Road surfaces on RAC rallies are a good test of any car's suspension and ride capabilities. Despite sporting a good-luck radiator mascot in the shape of an aeroplane, the car pictured has damaged its right-hand wing slightly.

Left SPF 777 seems intent on making the most of its 36-foot turning circle on this test. Driven by A. P. Warren, this car competed in the 1953 and 1954 Monte Carlo Rallies. Interested spectators include a Mk VII Jaguar on the left.

Below left The 1930s and 1950s were among the best years for the private car owner/enthusiast. Although factories entered teams in endurance events such as the Alpine Rally, Tulip Rally and Monte Carlo Rally, a well-prepared private entrant also stood a good chance of winning one of the class prizes. The Monte Carlo was of the most famous and taxing of the motoring events, and meant living in a car for three days and nights, and covering approximately 2,050 miles. Having survived the various difficulties of the mountain routes, the survivors arriving in Monte Carlo then had to cover five laps of the Monaco Grand Prix circuit.

The 1954 Monte Carlo Rally had eight starting points, Glasgow, Stockholm, Athens, Lisbon, Munich, Oslo, Palermo and Monte Carlo. Those starting from Glasgow had to rendezvous in Dover, as No 194, driven by J. G. B. Campbell and R. D. Barrack, is doing here.

Above Twelve Rileys were entered in the 1954 event among the 121 British entries. This travel-stained 2½ litre Riley was entered by J. W. Bowdage and J. E. Wright. Not every British competitor started from Glasgow - Stirling Moss, for example, started from Athens. Another well-known driver competing from the Munich start was Sheila Van Damm, whose father owned the famous 'We Never Close' Windmill Theatre in London.

Left L. O. Simms and A. P. O. Rogers pilot their Riley 2½ litre around Monte Carlo in 1954. It was not a great year for British entrants - Simms and Rogers gained 22nd place - but the Sunbeam-Talbot team of Stirling Moss, Leslie Johnson and Sheila Van Damm gained the Charles Faroux Challenge Trophy. The Rally winner, a Lancia driven by Chiron and Basadonna, was the subject of an eligibility row.

Below left The Riley of H. Feldman and J. Strang lined up for the Coachwork Competition. An extra spotlight can be seen above the windscreen, while extra spares are loaded on to the roof rack. Three of the prizes were concerned with road safety and equipment, one of which was the RAC Trophy.

Above Back home, we see an impressive mixture of entrants in the 1954 *Daily Express* Trophy at Silverstone. No 18 is a Riley 2½ litre, No 10 an Alvis TC21, No 2 a Jaguar Mk VII and Nos 11 and 14, waiting to take off on the left, are Daimlers.

Above Winner of the 1954 Redex Rally Concours d'Elegance Lady and Car competition was a Riley Roadster. In Concours competitions it was usual for the lady to dress to match the car, and sometimes it was difficult to decide which was wearing the most stunning coachwork!

Below The driver and passenger of this 2½ litre Roadster seem to be conferring during the 1952 Eastbourne Rally.

This Riley Roadster is taking part in one of the many driving tests set during the 1952 Welsh Rally.
The divided bumpers of this model were given extra chrome to appeal to the contemporary American market.

With sidescreens up, a 2½ litre Roadster shows its turning abilities on the 1954 Morecambe Rally. Note the unusual design of lamp-posts on the promenade.

Another promenade, this time at Blackpool, during the 1954 Blackpool Rally tests. The 2½ litre was a manoeuvrable car, helped by independent front suspension and rack-and-pinion steering. It was, however, a heavy car and hard work when handled at low speeds.

Left In May 1953 a 17-lap race for production touring saloons followed the main BRDC International Trophy meeting at Silverstone. Stirling Moss won the event in a Mark VII Jaguar, with a 2½ litre Riley driven by G. H. Grace second. The same Riley was first in Class D for cars of 2,000 to 3,000 cc, while the Riley team of G. H. Grace, A. P. O. Rogers and G. Gelberg won the team prize.

Below left A late 1940s Riley 2½ litre passing a line of more exotic machinery and an advertisement for *Motor Cycling* at Mallory Park.

Riley-based Healeys

Donald Healey had been a successful rally driver in the pre-war years; he was also Technical Director of Triumph Cars. The next stage in his career was to set up on his own as a sporting car manufacturer. His first creation was the Healey Westland Roadster, with a chassis manufactured by Westland Aeroparts and power supplied by a 2½ litre Riley engine. The curvy Elliott sports saloon proved to be the fastest production car in the world at the time of its introduction in 1946, and a 2.4 litre sportsmobile was produced in very small numbers. The two-seater Silverstone of 1949 was a potent sports-racing car, then in 1950 came the 2.4 Tickford saloon and Abbott Drophead coupé. Also produced were the Nash-Healey and the G-type with an Alvis engine. Later it became obvious to Donald Healey that supplies of Riley parts would eventually be no longer available, and he began his famous association with Austin-engined sports cars.

Above Lt-Col N. P. Burrows's Healey 2.4 litre Elliott saloon taking part in the 1952 *Daily Express* Rally. The car was later to win its class in the Concours d'Elegance for closed cars costing up to £2,000.

Above A Healey 2.4 litre convertible and lady taking part in the Concours d'Elegance Lady and Car competition. This event was organised by the BARC club as part of its Eastbourne Rally. Unusually the cars were indoors and driven into the middle of the Ballroom of the Winter Gardens. A footman would open the car door, the lady would step out, walk around, get back into the car, and be driven off.

Left The lightweight Healey Silverstone was an ideal competition car. It had much stiffer springing than the saloon cars, and top speed was in the region of 107 mph. It was rather like the pre-war Peugeot in that the head-lamps were placed behind the radiator grille. E. Ainsworth's Silverstone is competing in a trial organised by the Lancashire Automobile Club in 1950.

Above **2½ litre:** Some very fine artists were hired by most of the major motor manufacturers to bring out the best in their vehicles, as is evident in this rendition of the Riley 2½ litre.

Below **Pathfinder:** A front view showing the traditional Riley radiator grille. The bonnet and grille lifted as a single unit.

Below The colour options available for the August 1956 Riley range. There was a general fashion for duotone paintwork throughout the car world during this period; on the MG, for instance, this two-tone colour scheme was known as Varitone.

The Lively Riley *One-Point-Five*

Great performance in a compact four-seater.
Sports-tuned twin carburetter 1½ litre Riley engine.
Extra high top gear — that is virtually an
overdrive — for high speed cruising.
Acceleration is breath-taking.
Running costs are remarkably low.
Interior finish with real leather
upholstery and polished walnut
veneer. Yes! For performance,
comfort and value for money
it's a winner.

Two Brilliant Rileys for Magnificent Motoring

One of the most advanced cars you can
buy today. A luxurious, beautifully appointed
six-seater with graceful modern lines.
The powerful, six-cylinder engine gives
silky smooth running. Road holding and
cornering are outstanding.
Power-assisted brakes make fast
cruising effortless and safe.

*Three transmissions
to choose from.
A synchromesh gearbox
is standard, overdrive
or a completely
Automatic Transmission
are available
at extra cost.*

Every RILEY carries a
12 months' warranty and is
backed by Europe's most
comprehensive service — B.M.C.

The Magnificent Riley *Two-Point-Six*
with fully Automatic Transmission.

RILEY MOTORS LTD., *Sales Division:* COWLEY, OXFORD   BY.
London Showrooms: 55/56 Pall Mall, SW1. Overseas Division: Nuffield Exports Ltd., Oxford & 41-46 Piccadilly, W1

One-Point-Five and Two-Point-Six: This was the advertisement that Riley used to publicise these two models during 1958.
The BMC rosette is placed discreetly in the corner, alongside the statement offering a year's warranty. The Motor Show cars
of that year were red, and blue and white.

the ◆ *Riley* ◆ two-point-six

Two-Point-Six: This was not the most successful saloon to come out of the BMC stable - only 2,000 were made before the model was dropped in 1959. It was aimed at those who wanted a fast touring saloon, and to win over previous Riley owners and those who were aware of the Riley's reputation for having something 'extra' about it, BMC marketed the comparative luxury of the new model. The front cover of this 1958 brochure is clearly intended to impart a wealthy 'clubby' feel.

Right **4/72:** A colour chart of the choices available for the 4/72, Elf and Kestrel.

Below A Riley 4/72 series of 1961-69 in Arianca Beige/Sandy Beige.

FLORENTINE BLUE—BU. 7	DOVE GREY—GR. 28	SNOWBERRY WHITE—WT.4
OLD ENGLISH WHITE—WT. 3	SANDY BEIGE—BG. 7	AQUAMARINE—BU. 43
CUMBERLAND GREEN—GN. 35	ARIANCA BEIGE—BG. 13	AGATE RED—RD. 22
WHITEHALL BEIGE—BG. 4	PALE IVORY—YL. 1	TRAFALGAR BLUE—BU. 37
DAMASK RED—RD. 5	ALMOND GREEN—GN. 37	IRIS BLUE—BU. 12
BIRCH GREY—GR. 3	MAROON—RD. 23	SMOKE GREY—BU. 15
YUKON GREY—GR. 7	PORCELAIN GREEN—GN. 17	

Above Obviously designed to appeal to the lady motorist, the Riley Elf was advertised as 'Magnificent Motoring in Miniature', harking back to a time-honoured Riley advertising slogan.

Above **Elf:** Unlike the basic Mini, the Elf had a well-finished interior - the prospective buyer was asked to look at the carpet-lined interiors of the door pockets. The makers claimed to have equipped the car with a 'full range of driving instruments (more than in any other baby car)'.

Below **Kestrel:** Promoted as a sporting family car, the Kestrel of the late 1960s had an attractive front end treatment.

3. THE BMC YEARS

As part of the Nuffield Organisation's centralisation project, Riley production was moved from Coventry to Abingdon. The move started in 1948 and for a time it is possible that the cars were built at both places, although there are no records to prove this. Abingdon was chosen as Riley's new home as MG sports cars were already built there, and it seemed logical to have both sporting marques under the same roof. The Coventry factory would now build engines for several of the Nuffield cars, with Abingdon being the final assembly point.

Even more serious for Riley was the merger between Austin and Morris in 1952 to create the British Motor Corporation. In an attempt to ride out the difficult postwar years Austin and the Nuffield Organisation had worked together in an effort to keep both firms in business; in 1952 it was decided to make such action official, and the BMC Group was formed. This saved both companies a great deal of money in research development and gave them more buying power. At the time BMC was the largest car manufacturer in Europe and fourth largest in the world. Initially Lord Nuffield was Chairman with Leonard Lord as Deputy Chairman and Managing Director; Nuffield resigned after a few months leaving the Chair to Lord.

The prudence of the merger was borne out by the failure of old traditional marques like the Yorkshire carmaker Jowett - formed in 1906, the company did not survive beyond 1954 - and the Coventry firm of Singer, which became part of the Rootes Group in 1956. To begin with both Austin and Morris wanted to keep the identity of each individual marque separate, and in the early years little was done to alter this plan.

A newcomer was added to the Riley range for 1953. Known as the Pathfinder, it replaced the 2½ litre RMF. The new car still had the well-loved 2½ litre engine, but there the family resemblance ended. Outwardly the Pathfinder looked similar to the Wolseley 6-90, which was itself a larger version of the 4/44 series. One of the obvious differences between the Riley and the Wolseley was in the gear-change - the Wolseley's was mounted on the steering column, while that of the Riley was floor mounted on the driver's right-hand side.

The motoring press of the day was kind to the Pathfinder; *The Motor* described it as 'a comfortable and robust saloon car of exceptional performance', and went on to praise it as the most comfortable Riley made so far. Much was also made of its performance. However, despite these good press reviews, only 5,152 were built before the model was discontinued in 1957; the last link with the traditional Riley was now gone.

Perhaps it should be remembered that the basic price before tax of the Pathfinder in 1957 was £940, and would have had to compete with cars such as the Rover 75, Jaguar 2.4 and Humber Hawk. In the lower price ranges there were the Ford Zodiac at £645 and Morris Isis at £607.

It is perhaps likely that any car introduced to replace such an enthusiasts' car as the RM Series would be bound to encounter trouble. Although the RM Series was ageing by 1954, the dedicated Riley owner of the time would have been cheered by this quote from *The Motor* of 12 May 1954: 'There are 1½ litre cars that are livelier, or roomier, or more economical, but the Riley couples a happy mean in these respects with such a sterling blend of all that is in the best British traditions of good engineering, excellent finish, good looks and notably roadworthy behaviour that its appeal is not hard to understand.'

The next car to bear the Riley name was the Two-Point-Six sports saloon of 1958. This was the first time that a Riley shared all of its engine and component parts with another make of car, its BMC C-type engine being common with the Wolseley 6-90, although a minor difference was the Riley's 101 bhp as compared to 97 bhp of the 6-90. Although luxuriously appointed, and with several safety features, the car was not a success, only 2,000 coming off the production line. The steering was heavy, for example, making the car ponderous to turn.

Soon after the introduction of the Two-Point-Six came the smaller One-Point-Five, which was also produced as a Wolseley 1500. The latter may have had more luxurious appointments and more subtle colour schemes, but the Riley had twin SU carburettors and a top speed of 85 mph. Both cars had the benefit of the Morris Minor suspension and floorpan, and rack-and-pinion steering.

In many ways this was an ideal car for the late 1950s. Popular motoring was under way again after the shortages of the early post-war years, and more roads were being built; for example, the Preston Bypass, the first stretch of motorway to be built in Britain, was opened in 1958. It once more became the usual thing to take the car on family outings, and the Riley One-Point-Five was marketed to appeal to the family man who also liked a car with some performance, yet was not too expensive to run. A small car with a big 1,489 cc engine, it provided a tenuous link with Rileys of the past. The model also earned a degree of success as a competition car, doing surprisingly well in car club events.

The larger Riley was restyled for 1959 together with the rest of the BMC range, which now sported Italian styling by Pininfarina. The Riley models were known as the 4/68, later replaced by the 4/72, the other BMC versions being the Austin Cambridge, MG Magnette, Wolseley 15/60 (later replaced by the 16/60), and Morris Oxford. The Farina design also showed a family resemblance to the luxury Vanden Plas 3 and 4 litre.

'Magnificent Motoring in Miniature' arrived with the little Riley Elf of 1961. Probably best described as a 'Mini with a bustle', the Elf was a better finished and more expensive version of the famous BMC model. The marketing of the Elf seems largely to have been aimed at ladies, and it was promoted as a fun car. It is quite possible that it was to be seen as a second car, the major family car being the Riley 4/72, of course! Probably the closest rival of the Elf and Mini was the Hillman Imp of 1963.

Although small, the little Elf had good performance - there was even a tuned version capable of 110 mph - and shared the Mini's low petrol consumption figures. A Riley Elf took part in the annual Mobil Economy Run of 1965 and won the class for cars of 500 to 1,000 cc with a consumption of 54.43 mpg.

In some ways it is perhaps fitting that the last car to be built under the Riley name should be christened 'Kestrel'. Although a fair number of Riley traditionalists were upset by the use of this name on a car of the BMC 1100/1300 range, it was comfortable, enjoyed good performance and was a cut above the Austin/Morris range. Many people bemoaned BMC's 'badge engineering', but it was a common feature of the era: Rootes, for instance, produced 'badge engineered' Hillmans, Singers and Sunbeams, and Rolls-Royce also manufactured Bentley versions.

The 'life of Riley' was cut short in 1969 with the creation of British Leyland from BMC and Leyland Motors, possibly because the marque was not produced in great enough numbers. During its 70-year lifetime Riley had seen car production move from Coventry to Abingdon, to Cowley and finally Cowley and Longbridge. The Morris name was to disappear in 1984, followed by Austin in 1988; Wolseley had already followed Riley in 1975. Of these names it is only MG that remains.

The Pathfinder

The styling of the Pathfinder brochure leaves us in no doubt that BMC intended the model to be aimed exactly at the sort of people who bought the RMF; while the RME was still in production the two cars were advertised side by side. With an overall length of 15 ft 3 in and a width of 5 ft 7 in, the Pathfinder was a large car weighing 2,800 lbs.

Introduced in the Coronation year of 1953, the Pathfinder was the last car to bear any resemblance to the traditional Riley. The engine was the well-tried favourite, the four-cylinder 2,443 cc twin-cam unit. Gerald Palmer, who had previously worked for Jowett, was in charge of the new designs for Riley, Wolseley and later MG, and while the radiator grille was designed in the old Riley style, the rest was an entirely different design.

SENSIBLE GLOVE BOX. The capacious glove box on the passenger side has a lid which opens downwards, making a useful tray and preventing the contents from falling out.

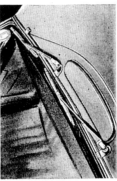

DUAL WINDSHIELD WIPERS Driving vision is always clear and safe as twin wipers sweep the windshield in wide arcs. They are self-parking, too! Windshield washing equipment is also a standard fitment.

SAFE, CONFIDENT REVERSING Rear window is extra large and curved to blend with the contour of the body. The driver has an excellent rearward view. The 13-gallon (59-litre) fuel tank is fitted with a locking filler cap.

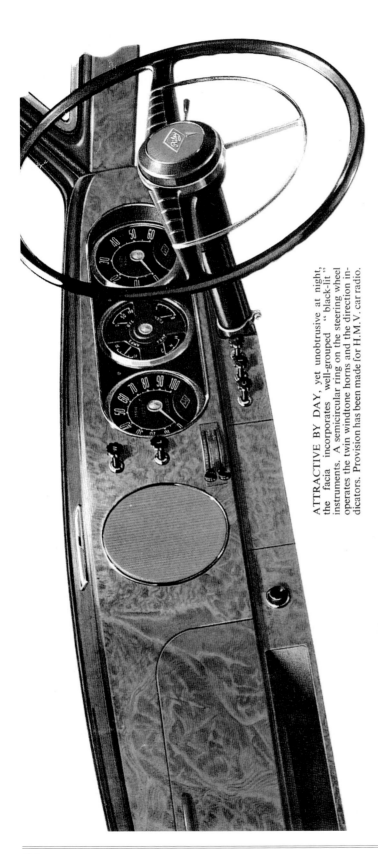

ATTRACTIVE BY DAY, yet unobtrusive at night, the facia incorporates well-grouped "black-lit" instruments. A semicircular ring on the steering wheel operates the twin windtone horns and the direction indicators. Provision has been made for H.M.V. car radio.

The fascia of the 1956 Pathfinder. The finishings were different from those of its Wolseley 6-90 counterpart, but similar in quality to the MG. Intended as a performance saloon, the Pathfinder had a rev-counter fitted as standard, and provision was made for an HMV car radio if required. It is easy to think of features like windscreen washers as being a fairly recent innovation, but they were standard on the Pathfinder of this time.

Above A three-quarter view of the new Pathfinder, which was priced at £1,240 with Purchase Tax. The name was written in script at the bottom of the wing. For better ease of access to the engine, the bonnet and radiator shell lifted up as one unit.

Right As an entirely new design the Pathfinder was given a new box-section chassis. The gearbox was also unusual in that the gear lever was placed on the driver's right-hand side, which allowed for the provision of bench seats in the front. Rear suspension was controlled by hydraulic dampers, front by 'Torsionic' independent front suspension, controlled by telescopic hydraulic dampers. A Borg-Warner Overdrive unit was an optional extra.

BMC dropped the Pathfinder in 1957. Although comfortable and a good performer, the car's price and petrol consumption of between 17 and 26 mpg counted against it. There were by now many other cars of similar specification produced by other car-makers such as Ford.

Two Riley Pathfinders taking part in the 75-mile, 25-lap Production Touring Car race at Silverstone in May 1955. Although Jaguar carried off the honours, Riley cars came second and third in the 2,001-3000 cc Class. The Pathfinder driven by A. P. O. Rogers blew a big-end and had to retire.

Above KRX 820 again, this time at Oulton Park in October 1957. Pathfinders had a top speed of 97 mph and a 0-60 mph acceleration time of 18.8 seconds.

The Two-Point-Six

Right The replacement for the Pathfinder was the Riley Two-Point-Six. Although similar in appearance to its predecessor, this design was purely a BMC Series III Wolseley 6-90 apart from the badge. Power came from a six-cylinder 2,639 cc BMC C-series engine, with 101 bhp rather than the Wolseley's 97. It boasted power-assisted brakes, and was also the first Riley model to be offered with an automatic gearbox as an option.

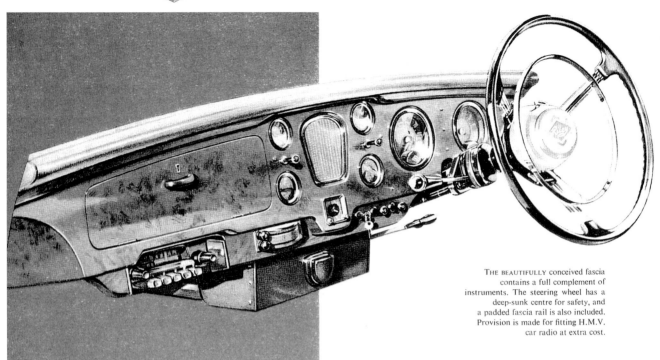

THE BEAUTIFULLY conceived fascia contains a full complement of instruments. The steering wheel has a deep-sunk centre for safety, and a padded fascia rail is also included. Provision is made for fitting H.M.V. car radio at extra cost.

Left The fascia of the Two-Point-Six. This time the radio was fitted as standard, and for the first time safety was made part of the advertising campaign. The dashboard was padded, and there were references to the steering wheel having a deep-sunk centre for safety. However, one of the car's road testers was not very complimentary about the steering wheel arrangement: he said that the horn ring 'rattled annoyingly'.

Below A handsome car, the Two-Point-Six was available in a choice of colours: Black with two-tone green, two-tone maroon, or biscuit and brown upholstery; Basilica Blue and Florentine Blue with two-tone grey upholstery; Shannon Green and Leaf Green with two-tone green or biscuit and brown upholstery; Yukon Grey and Birch Grey with two-tone grey or two-tone maroon upholstery; Maroon and Kashmir Beige with two-tone maroon upholstery (pictured); and Black and Birch Grey with two-tone maroon or two-tone grey upholstery. Monotone finish was an option in Black, Basilica Blue, Shannon Green, Maroon or Yukon Grey.

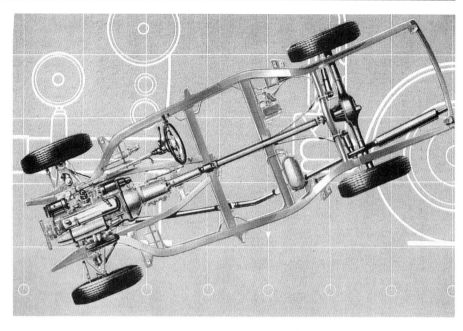

Above The chassis of the Two-Point-Six was of welded box-section constructed of steel. Although sturdy, it was less than comfortable over rough roads.

During its short life the Riley Two-Point-Six was produced at Abingdon and later at Cowley. Production ended in 1959, and the next larger Riley models were totally different in appearance and were designed by Pininfarina.

The One-Point-Five

Left and above Soon after the launch of the Two-Point-Six came the compact Riley One-Point-Five. Introduced first in Wolseley 1500 form, the One-Point-Five came into being towards the end of 1957. Alec Issigonis had set new standards for small family cars when his Morris Minor had been unveiled at the 1948 Motor Show, the steering and suspension earning much respect, and the Wolseley 1500 and Riley One-Point-Five both shared the Morris Minor's floorpan, suspension and steering. In style the One-Point-Five was supposed to be a small sporting saloon, and had very similar lines to the Austin A55. Manufactured for eight years with few alterations, it was marketed as the smaller Riley that thousands had been waiting for. Some 39,600 were manufactured.

You'll take your hat off to the

NEW

Riley

One-Point-Five

powered for thrilling performance

PUBLICATION No. H. & E. 5980

The front cover of the sales brochure for the 1959 Riley One-Point-Five shows the car's ideal driver! The Riley was marketed as the sports version of the Wolseley 1500, the latter having an altogether different campaign. According to One-Point-Five literature, 'Steering and roadholding are superb even by Riley standards. The sports-tuned, twin-carburettor engine is matched with a specially high top gear, which is actually better than overdrive.' It had a four-cylinder 1,489 cc engine, with a power output of 68 bhp, which compared favourably with the 50 bhp performance of the Wolseley 1500. Although said to designed with the enthusiast driver in mind, it was probably bought by the family man who had left

his sports car days behind. Priced at £816 in 1959, including Purchase Tax, the car was good value for money.

Although not so luxuriously appointed as the Wolseley, the 1.5 had nice touches of its own. There was a polished wood fascia, and a rev-counter was included in the instrumentation. The choice of colour schemes was Birch Grey, Damask Red, Florentine Blue, Leaf Green, Yukon Grey, and Black. In two-tone, which always made it look a bit squatter, but was supposed to make it look sportier, there was a choice of Damask red and Kashmir Beige; Yukon Grey and Birch Grey; Black and Chartreuse Yellow; Old English White and Leaf Green; Old English White and Florentine Blue.

Polished wood facia

The luxury facia panel of the Riley One-Point-Five is faced with beautifully grained polished walnut veneer.

Instruments include rev. counter

The instrument arrangement has classic simplicity with three matched large–diameter round dials. These include a rev. counter—a feature specially sought after by enthusiasts.

You'll take your hat off to these big features too!

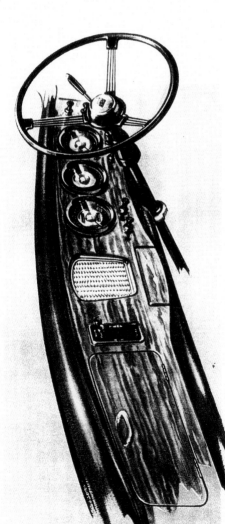

Two-colour luxury

Really comfortable seating for four adults. That's the designers' aim in the Riley One-Point-Five. Here you can see how liberally this aim has been interpreted. The interior has a striking contemporary smartness aided by the two-colour leather covering of the seats. Top-quality trim all round and polished walnut veneered cappings to the door panels complete the luxury effect.

The One-Point-Five received some minor changes in 1960, and again in 1961, by which time it was known as the Riley One-Point-Five Series III. The most obvious bodywork change was the grille below the radiator, which was given more horizontal bars. The price new was £799, including Purchase Tax.

The last year of production for the model was 1965. It had proved to be a good seller for BMC, perhaps not quite in the mould of tradition- al Rileys, but nevertheless winning friends and proving to be useful at car club events. The One-Point-Five makes a practical classic car today.

Above Surprisingly, or perhaps not so, the One-Point-Five proved a capable performer in competition events. This one is taking part in the 1957 Eastbourne Rally - its handling, together with a wheelbase of only 12 ft 9¾ in, made it ideal for manoeuvring in small spaces.

Below One firm that specialised in modifying cars to fit the needs of competition car drivers was Boon & Porter of Barnes, West London.

By 1958 the firm had been involved with tuning and preparing Riley cars for nearly 60 years; for the One-Point-Five they fitted an extra fuel tank in the boot on the nearside, an anti-roll bar at the front, a fly-off type handbrake, Lucas fog and spot lamps, and a two-note horn. Rally drivers had the benefit of an extra parcel tray and a concealed map reading light. Cylinder-head modifications were also offered.

The Riley One-Point-Five's B Series four-cylinder 1,489 cc gave it a top speed of 85 mph. This example is leading a Volvo 122 (Amazon) at Silverstone in 1959. . .

. . . while No 100 is taking part in a sports car event at Brands Hatch in the same year. The belt holding down the bonnet at the front would seem to suggest that the car has been race modified.

C. P. Wiggins partnering a One-Point-Five at a race meeting at Silverstone in May 1961.

The 4/68

Above 1959 was a benchmark in British motoring history. It was the year that the M1 was officially opened, and the year of the Mini, the revolutionary small front-wheel-drive car designed by Alec Issigonis. Less importantly, BMC's Farina-styled B Series model range of large, 'boxy' cars with styled-down tail fins was launched. The Riley version was known as the 4/68, and its companions were the Austin A55, Wolseley 15-60, MG Magnette Mk III, and Morris Oxford Series V.

Right The front cover of a 1961 4/68 sales brochure. This was to be the model's last year of production, 10,940 cars having been built during a production run of three years.

if you're **GOING AHEAD**, go with a *Riley*

4 *SIXTY EIGHT*

Take a
good
close look
at the

Riley

Explore the many advanced features and extra refinements that the two latest Rileys offer. Check for performance . . . for quality of workmanship . . . for comfort . . . for style. Be as critical as you like — we think you'll be convinced that for sheer excellence and outstanding value-for-money these Rileys are, each in its class, your very best buy.

The beauty of the Riley 4 Sixty Eight is that it is many good cars in one—a sports car—a business car—a family car—built to give you year after year of magnificent motoring. Price: £725 plus £303.4s.2d. P.T. Duotone colours extra.

The lively Riley One-Point-Five is a compact four-seater with big-car performance. Extra high top gear for carefree high speed cruising. Interior finish shows British workmanship at its best. Price £575 plus £240.14s.2d. P.T.

Riley

Every Riley carries a 12 months' warranty, backed by Europe's most comprehensive service—B.M.C.

RILEY MOTORS LTD., *Sales Division, Cowley, Oxford. London Showrooms: 8-10 North Audley St., W.1. Overseas Division: Nuffield Exports Ltd., Oxford and 41-46 Piccadilly, W.1.*
R116

This 1960 advertisement for the Riley range shows the difference
between the earlier more rounded styling and the new shape of the 4/68.

Right Many of the early 1960s BMC cars were lacking in individuality, but like the MG Magnette the Riley 4/68 was marketed as a luxurious sports saloon. Wessex Motors of Salisbury carried out their own conversions on the 4/68 and called it the Riley Riviera. The rear end of the car was re-designed with the tail fins cut down and different rear lamps fitted. The car was re-sprayed royal blue and wire wheels were fitted. Mechanically the major change was the fitting of a 1,588 cc MGA engine and clutch, which helped the car to attain a top speed of nearly 100 mph. In all, three stages of conversion were offered at varying prices.

Below The fascia of the 4/68. That the interior still combined Riley performance with Wolseley standards of finish is shown by the retention of the rev-counter on the instrument panel.

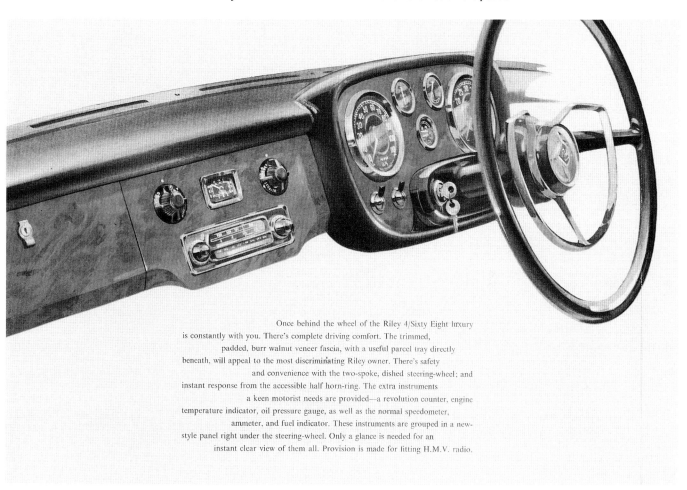

Once behind the wheel of the Riley 4/Sixty Eight luxury is constantly with you. There's complete driving comfort. The trimmed, padded, burr walnut veneer fascia, with a useful parcel tray directly beneath, will appeal to the most discriminating Riley owner. There's safety and convenience with the two-spoke, dished steering-wheel; and instant response from the accessible half horn-ring. The extra instruments a keen motorist needs are provided—a revolution counter, engine temperature indicator, oil pressure gauge, as well as the normal speedometer, ammeter, and fuel indicator. These instruments are grouped in a new-style panel right under the steering-wheel. Only a glance is needed for an instant clear view of them all. Provision is made for fitting H.M.V. radio.

The fact that the 4/68 was obviously intended as one of the first of the motorway 'cruisers' is shown by the positioning of the M1 motorway in the top right of the picture (*enlargement above right*). The 4/68 was available in a choice of Black, Porcelain Green, Old English White, Smoke Grey, Clipper Blue, and Maroon in single tone, and Black and Old English White, Maroon and Whitehall Beige, Clipper Blue and Smoke Grey, or Smoke Grey and Old English White in duotone colours.

The 4/72

Above The successor to the 4/68 was the Riley 4/72. In an incongruous effort to recall the Rileys of earlier days, the catalogue says, 'Like the Kestrel of the 1930s, it provides a civilised and amenable mode of transport with sporting freshness and performance.'

Above right and right Two views of the 4/72 taken at the Earls Court Motor Show in 1963, the year of Lord Nuffield's death. Introduced in 1961 the 4/72 was an improvement over the earlier model. Powered by a four-cylinder 1,622 cc 68 bhp engine, top speed was 86 mph. Although visually identical to the 4/68, the 4/72 had anti-roll bars fitted, a wider track and a longer wheelbase. This resulted in much better handling.

The Riley 4/72 had a fairly long production run at BMC's Cowley works, spanning the years 1961-69, and was the last of the larger Rileys to be manufactured. This is a 1967 example. Priced at £921, the 4/72 was the most expensive car in the comparable BMC range: the MG Magnette cost £900, the Wolseley 16-60 £853 the Morris Oxford £747, and, last but not least, the Austin A60 Cambridge at £737.

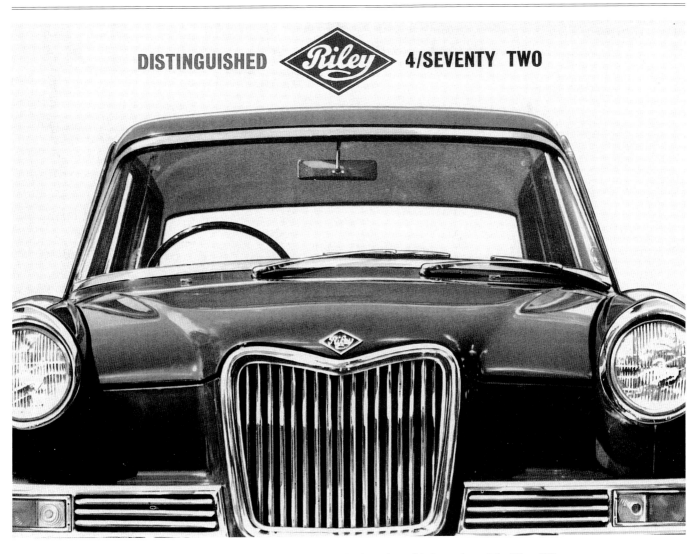

DISTINGUISHED *Riley* 4/SEVENTY TWO

Above Distinguished and dignified are words describing this front view of the Riley 4/72.

Right A close-up of the steering wheel, illustrating the neat automatic transmission controls.

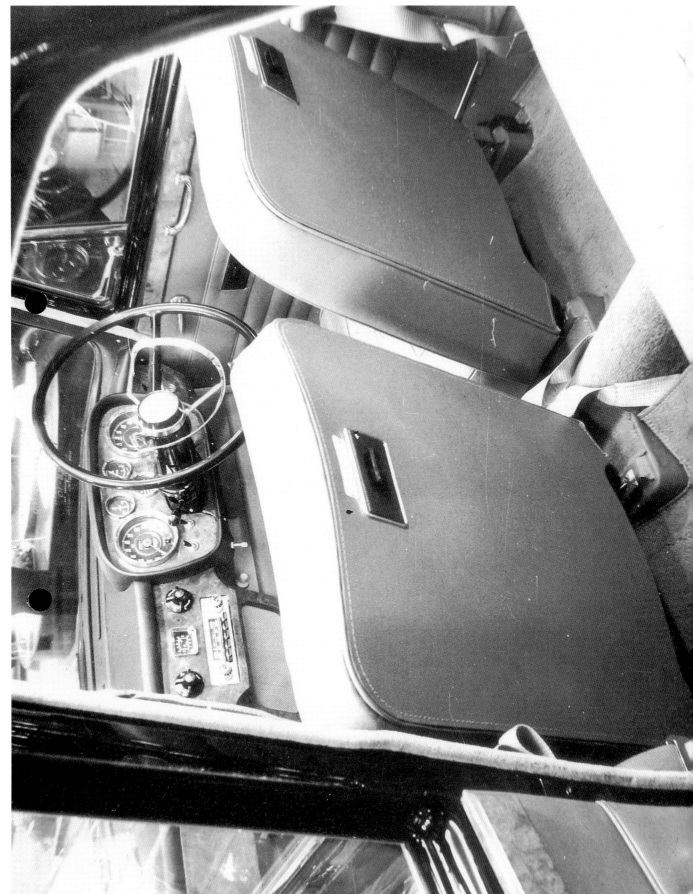

The interior of the 4/72, showing (*above*) the driving position, the neat instrument panel and the floor-mounted dip-switch. This car is fitted with the optional automatic gearbox. Among the interior refinements (*below*) are twin ash-trays built into the rear of the front seats. Although it would be some years before the wearing of seat belts became compulsory, it is interesting to note that they are fitted to this show car; seat belt anchorage points were standard on all BMC cars from 1961.

The Elf

SHE'S NEW...

Above In 1961 BMC introduced its booted Mini, the Riley Elf, also referred to as the 'Mini with a bustle'. This side view from the Elf sales catalogue shows how like the Mini the car really was, apart from its 'bustle', of course, and like the Mini the Elf was a true four-seater.

Below A view of the Riley Elf 'bustle' at the 1961 Earls Court Motor Show.

Above The little Riley was not in production for long before its 848 cc engine was replaced by a new 998 cc unit in October 1962. Only a few of the earlier Elfs were manufactured, so they are comparatively rare today.

Right The Wolseley Hornet was introduced at the same time, and both cars were launched at the 1961 show. The obvious differences were in the grille designs and bumpers.

At Earls Court in 1963 the Riley Elf keeps company with the likes of Singer, Skoda, Sabre and Saab, of which only the Skoda and Saab names survive today. From 1963 until 1966 the car was known as the Riley Elf Mk II. There were few obvious physical changes but the 1963 car pictured here has gained a Riley blue diamond and over-riders on the rear bumper.

At £693 19s the Elf was dearer than the Mini, but had the benefit of a heater, screen-washers and more luggage space, making it more suitable than the Mini for small families. The original 848 cc engine produced 34 bhp, and top speed was about 70 mph. It shared many of the Mini's personality traits, and was a popular small car.

Below The front view of the Elf, the radiator grille still bearing the Riley blue diamond and showing some resemblance to the old style radiator. With the 998 cc engine the Elf could manage 0-60 mph in 24.1 seconds and had a comfortable cruising speed of 62 mph with a top speed of 70 mph.

Above right The front cover of the 1962 Elf sales brochure is obviously promoting the Elf's smartness and appeal as a shopping car. As with the Mini, the Elf was ideal for parking in small spaces with a turning circle of 31 ft 7 in. There were six colour schemes available: Chartreuse Yellow with Florentine Blue roof; Cumberland Green with Old English White roof; Damask Red with Whitehall Beige roof; Florentine Blue with Old English White roof; Birch Grey with Old English White roof; and Yukon Grey with Birch Grey roof.

Right The last Riley Elf was the Mk III, which ran from 1966 until 1969, when the Riley marque itself was discontinued. This version had the luxury of wind-up windows.

Above Another nice luxurious touch to the Elf Mk III was leather seats.

Left Issigonis also designed the Austin/Morris 1100 series, which entered production in 1963 and were intended to be larger versions of the Mini. One of the design features of this new series was 'hydrolastic' suspension, which new innovation also featured in the Mini, Riley Elf and Wolseley Hornet. It was not entirely successful, because as the cars got older some of them developed 'saggy' suspension. Automatic transmission was available as an option for those who really needed it on a small car.

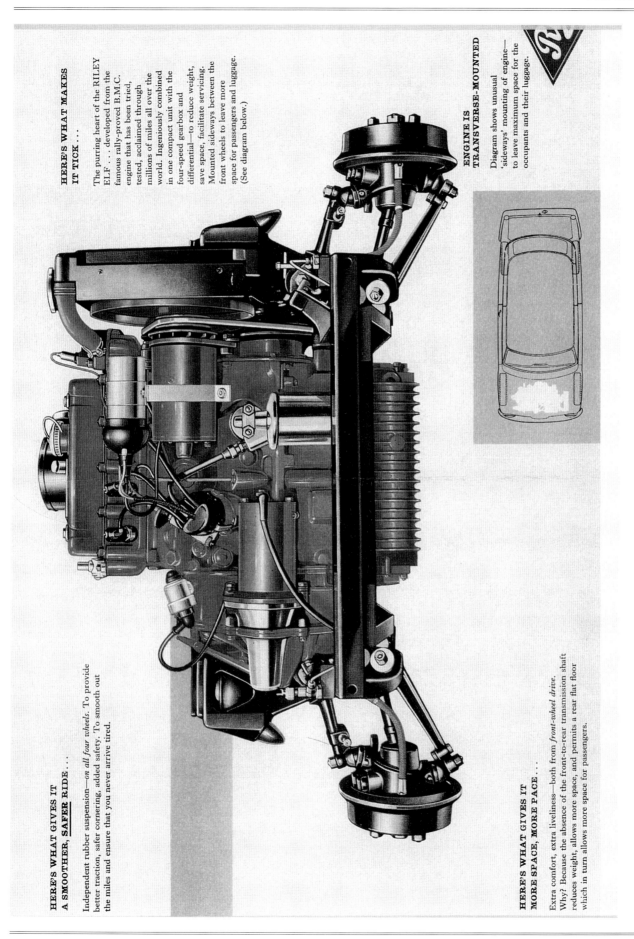

HERE'S WHAT MAKES IT TICK . . .

The purring heart of the RILEY ELF . . . developed from the famous rally-proved B.M.C. engine that has been tried, tested, acclaimed through millions of miles all over the world. Ingeniously combined in one compact unit with the four-speed gearbox and differential—to reduce weight, save space, facilitate servicing. Mounted sideways between the front wheels to leave more space for passengers and luggage. (See diagram below.)

ENGINE IS TRANSVERSE-MOUNTED

Diagram shows unusual 'sideways' mounting of engine— to leave maximum space for the occupants and their luggage.

HERE'S WHAT GIVES IT A SMOOTHER, SAFER RIDE . . .

Independent rubber suspension—*on all four wheels*. To provide better traction, safer cornering, added safety. To smooth out the miles and ensure that you never arrive tired.

HERE'S WHAT GIVES IT MORE SPACE, MORE PACE . . .

Extra comfort, extra liveliness—both from *front-wheel drive*. Why? Because the absence of the front-to-rear transmission shaft reduces weight, allows more space, and permits a rear flat floor which in turn allows more space for passengers.

Another illustration from an Elf sales brochure, showing how truly Mini-based the little car was. The Mini became a favourite rally car, and soon successfully tuned versions such as the Mini Cooper and Mini Cooper S were in production. Less well-known, however, was a tuned Riley Elf, which could be obtained from Speedwell Performance Cars in Finchley Road, North London. The engine of this 'Executive' Riley Elf was bored out to 67.1 mm, and a Mini-Cooper crankshaft was fitted to give a stroke of 81.3 mm, giving an engine capacity of 1,150 cc. This exercise combined with tuning gave a power output of 90 bhp. Top speed of the Speedwell Riley was about 110 mph, but at that speed the ride was described as 'a little lively'.

The Kestrel

FOR MAGNIFICENT MOTORING

ELF · KESTREL · 4/SEVENTY TWO

The 1965 range of Rileys, small, medium and large, was completed with the arrival of the Kestrel. Several enthusiasts were dismayed by the company decision to use such a well-loved name again, this time on one of many of the BMC 'badge engineered' 1100s.

495 CJB

Street life in Great Britain during the mid to late 1960s must have been slightly confusing for anyone with only a passing interest in the motor car. In the BMC 1100/1300 range alone there was the possibility of encountering 15 different versions if you counted both the superseded and current versions. The Kestrel was, as before, promoted as a sporting family car.

Riley Kestrel

A full range of instruments is set off by the warm gleam of polished woods. Here, as elsewhere, quiet luxury is the keynote.

The most

FOR MAGNIFICENT MOTORING

Small-car handling and parking.
Big-car comfort for a family of five.
And all their luggage.

compact, most luxurious sporting 5-seater

As ahead of its time as its legendary pre-war namesake. That's the revolutionary new RILEY KESTREL. Setting the standard for the 1970s. Combining *performance* with *roominess* with *compactness* with *luxury* to a degree unmatched by any other sporting saloon of its size. The feel of real leather. The gleam of real wood. All reflect the character and craftsmanship that are so much a part of the Riley tradition. All contribute to the pride you'll feel when driving this new RILEY KESTREL.

Above left and above The Kestrel was very closely linked with the MG 1100 and shared the same running gear. This sales brochure declares that the 'revolutionary new' BMC derivative was 'setting the standard for the 1970s', but as we now know that was not to be the case. Power for the Kestrel was provided by a four-cylinder 1,098 cc engine unit with twin carburettors and a top speed of 85 mph. The price new, including Purchase Tax, was £781.

Colour options for 1965 were Snowberry White, Sandy Beige, Aquamarine, Agate Red, Cumberland Green, and Black. Duotones were available as Snowberry White/Sandy Beige, Snowberry White/Cumberland Green, Sandy Beige/Arianca Beige, and Aquamarine/Snowberry White.

Left The fascia of the Kestrel, and a new look to the instruments. It stole a march on the Wolseley and MG, which were still equipped with strip-style panels, but otherwise will be familiar to any former 1100/1300 owner. In keeping with the Riley tradition, there is a rev-counter.

the appeal of a space-saving weight-saving east-west engine

plus front-wheel drive for better road-holding, twin carbs. for added thrust

The lively twin-carburetter 1098 c.c. engine is combined with clutch, gearbox, and final drive in a single space-saving, weight-saving power pack. Then transverse-mounted to leave still more room for passengers and their luggage. Front-wheel drive means improved road-holding, further weight-saving, still better performance.

the added safety of front disc brakes

Self-adjusting disc brakes of 8″ diameter on front wheels ensure fade-free braking at all speeds. Pressure-limiting valve between master-cylinder and rear drum brakes apportions braking more accurately between front and rear wheels—for both wet and dry road conditions.

Above and above right A cutaway illustration showing the front-wheel drive and transverse engine unit. 'Hydrolastic' suspension was an important feature of the Issigonis-designed cars, and this diagram explains how it works. The claim of no maintenance proved over-optimistic, as many cars developed 'saggy' suspension in their later years. Some service stations offered to recharge the suspension system to bring the car back on to an even keel.

Right The whole range received a minor facelift during 1967, one of the most noticeable changes to the Riley being the fitting of side indicators. A more powerful 1275 cc engine was also fitted to the Kestrel in that year, and for a very short time this model was known as the Kestrel 1275. An 1100 version was also produced at the same time, but was phased out at the beginning of 1968.

The last Riley of all was the 1300 Mk II of October 1968, which had an uprated 70 bhp engine; the 'Kestrel' part of the name had been dropped earlier in the year. This model survived until July 1969, when British Leyland discontinued the Riley marque, ending 70 years of 'Magnificent Motoring'.

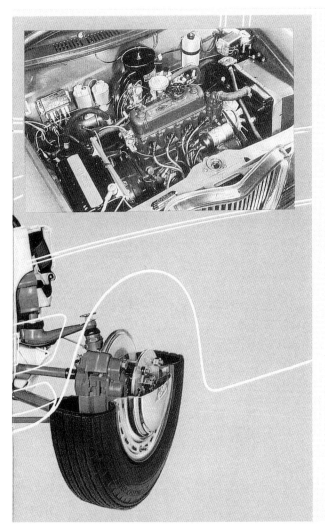

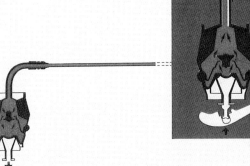

the supreme comfort
of hydrolastic® suspension . . .

to smooth the miles, absorb the hours and make
every journey more relaxing

No pitching. No rolling. No jolting. Because of separate pitch and roll control bars. Because of rubber-mounted sub-frames. Because, above all, of the luxury of 'Hydrolastic' suspension—the world's most advanced system of automobile springing.

This uses a water-based, anti-freeze fluid to apply the dynamic force to the rubber spring units on which each wheel is carried.

The fluid is interconnected, by pipes, to the front and rear suspension units on the same side of the car. This enables the wheels at the back to anticipate the action of those at the front. Thus road-bumps are ironed-out—and the Riley Kestrel maintains an even keel. Always.

That's 'Hydrolastic' suspension. No moving parts. No glands to leak or wear. No maintenance.

INDEX

Autovia 10, 23

British Leyland 62, 98
British Motor Corporation 61ff

Citroen Traction Avant 25, 44
Coachbuilders and tuners,
    Bonallack 32; Boon & Porter 75;
    Epps Brothers 42; Speedwell
    Performance Cars 93; Wessex
    Motors 79

Dixon, Freddie 16

Eyston, George 9, 15

Hawthorn, Mike 19
Healey, Donald 9, 55-6
'Hydrolastic' suspension 92, 98

Issigonis, Alec 71, 77, 92

James, Jimmy Ltd 10

Lord, Leonard 61

Mays, Raymond 17
Morris Motors 10
Motor Cycle Club 9, 12-3
Motor racing circuits, Brands Hatch
    76; Silverstone 49, 54, 66, 76

Motor Show, Earls Court 71, 81, 86,
    88

Nuffield, Lord 10, 61, 80
Nuffield Organisation 25, 33, 61

Palmer, Gerald 63
Parry-Thomas, J. G. 9, 15
Pininfarina 70

Railton, Reid 9
Rallies, Blackpool 53; Eastbourne
    43, 50, 75; Monte Carlo 46, 47, 48;
    Morecambe 52; RAC 19, 45;
    Redex 44, 50; Welsh 21, 51

Riley, Percy 7, 8; Stanley 7, 8, 9;
    Victor 7, 10; William 7

Riley models:
1½ litre (RMA) 10, 25, 26, 28, 31, 32,
    33, 34, 43; estate 32
2½ litre (RMB) 25, 26, 28, 30, 31, 35,
    36, 37, 38, 40, 44, 49, 53, 54, 57
4/68 62, 77-80
4/72 59, 80-5
12/6, 14/6 & 15/6 9, 11
Drophead Coupé 27, 29, 32, 43, 45
Elf 60, 62, 86-93
Forecar (1903) 7, 8
Imp 9, 19

Kestrel (1930s) 10, 15, 21
Kestrel (1965) 60, 94-99
Lynx 21
Monaco 9, 13
MPH 10, 19
Nine 9, 10, 12, 17
    Brooklands 15, 16, 18
One-Point-Five 58, 61-2, 70-6
Pathfinder 40, 57, 61, 62-7
RME 31, 38, 39, 40, 62
RMF 38
Roadster (2½ litre) 41, 42, 50, 51, 52
Sprite 20
Two-Point-Six 58, 59, 61, 67-70
Victor 10

Riley mascots 23
Riley Motor Club 14
Riley Specials, Healey 55-6; 'Red
    Mongrel' 16; 'White Riley' 17
Rush, Harry 25

'Torsionic' suspension 25, 44, 65
Trials and hill-climbs 9, 12-3, 17,
    20, 21, 22

Wolseley Hornet 87
Women's Association of Sporting
    Amateurs 22